YOUR KNOWLEDGE HAS VALUE

- We will publish your bachelor's and
  master's thesis, essays and papers

- Your own eBook and book -
  sold worldwide in all relevant shops

- Earn money with each sale

Upload your text at www.GRIN.com
and publish for free

Bibliographic information published by the German National Library:

The German National Library lists this publication in the National Bibliography; detailed bibliographic data are available on the Internet at http://dnb.dnb.de .

Imprint:

Copyright © 2016 GRIN Verlag, Open Publishing GmbH
Print and binding: Books on Demand GmbH, Norderstedt Germany
ISBN: 9783668319264

Martin Rakowitsch, Maximilian Scheid

Aus der Reihe: e-fellows.net stipendiaten-wissen

e-fellows.net (Hrsg.)

Band 2146

Experimental modal analysis of an automotive drive-train subframe

GRIN Publishing

SD2150 Experimental Structure Dynamics - Project Report

Martin Rakowitsch   Maximilian Scheid

May 31, 2016

Abstract: Within the framework of this project, the dynamic behaviour of a sub frame supporting the rear differential of vehicles is investigated. The original structure, consisting of sub frame and assembled gear, used hard plastic bushings. This setup caused noise inside the passenger cabin, which was not tolerated by GKN's customers. The emergence of noise was reduced by using soft elastomer bushings, but the causes are still unknown. In order to obtain a better understanding of the noise's source, an experimental modal analysis of the structure is performed. After draping a point cloud over the structure, the frequency response functions between these points are measured. The complex exponential method, including Prony's method, provides algorithms which are used to extract the structure's modes. Within the frequency range of approx. 50 to 550 Hz seven eigenmodes are extracted. Two of them show strong interaction between sub frame and differential, while the residual modes are dominated by the frame's mode shapes.

Contents

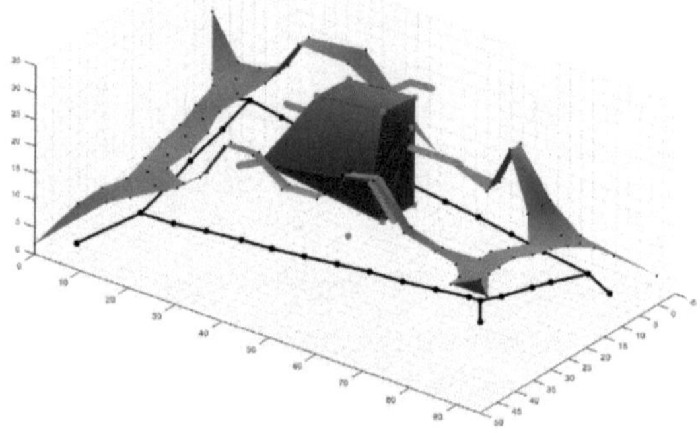

List of Figures

1   Introduction

Within the framework of this project, it is the aim to investigate the dynamic properties of an automotive sub-structure provided by GKN Driveline. The modal parameters are extracted experimentally and these results are used to animate the structure's eigenmodes and to hereby get a better understanding of its dynamic behaviour.

This section shall give a short introduction to the test object and the given problem. It is followed by a listing of the laboratory setup. Section 3 will present an explanation of the signal analysis and hereafter the principal theory behind mode extraction from the obtained frequency response functions (FRFs) is given. In the end, we will present our gathered results and use a self-designed MATLAB program to animate the eigenmodes of the structure. In our last section, we will discuss our findings, prove consistency and try to explain particularities.

Figure 1: Test structure.

Figure 1 depicts the given automotive subframe of GKN Driveline, which is the object of investigation. It features a basically rectangular metal hollow profile with a rectangular cross section. At its corners, it offers mounting points including an elastomer decoupling of its environment. In its centre, it carries a power train's transmission block, which is mounted by using soft bushings. Formerly used hard bushings caused significant noise in the passenger cell, but GKN assumes that the new soft bushings solved the problem. It is now the aim to understand how the soft bushings changed the overall dynamic behaviour of the structure. Therefore, the overall experimental setup is marked by the three following versions, which then have to be be compared.

- Assembled structure with soft bushings

- Assembled structure with hard bushings

- Frame without gearbox

It is the task of this project group to investigate the first configuration. The frequency range of interest is from 80 to 500 Hertz.

2   Laboratory setup

It is of vital importance to avoid unwanted interference of support loads when investigating the eigenmodes of a test object. Hence, we suspend the frame from the ceiling with soft elastic springs. Since our frequency range of interest is 80 - 500 Hz, it is recommended to make sure that the rigid body modes of the supported structure are at maximum 30% of 80 Hz to make sure that the support forces do not interfere with the elastic modes [1]. A short measurement of the period time of the oscillating up and down movement reveals a rigid body motion at 20 Hz. Later we will see that the elastic modes will begin above 100 Hz, so that we can neglect the support loads in our investigation. We need to overlay the test structure with a number of

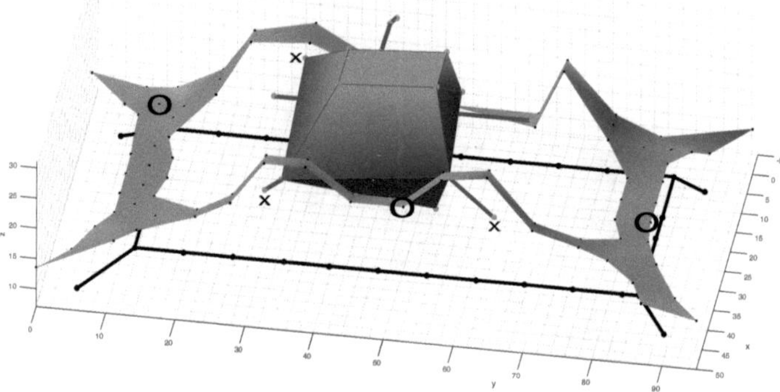

Figure 2: Grid points for the measurements.

measurement points. One have to find a good trade-off between accuracy and number of points. We pursue several strategies to extract the mode shape details that we would like to highlight hereafter. To detect relative motion of the gear box, we choose the connection points of the gear box and frame as well as the outer mounting points (black crosses in Fig. 2). Several points are located on the gear box to detect pitching and dipping. We expect an overall better visualisation of the bending modes by adding an array of points at the bottom of the frame as it is more flat. Additionally, it adds a backup set of points to filter for possible outliers. The deformation of the cross section requires us to put at least two points in transverse direction. For one side we even decided to add a third row (cf. Fig. 2 left side and 3). The overall amount of grid points adds up to 134 nodes. Additionally, we defined a triangulation table similar to the mesh generation

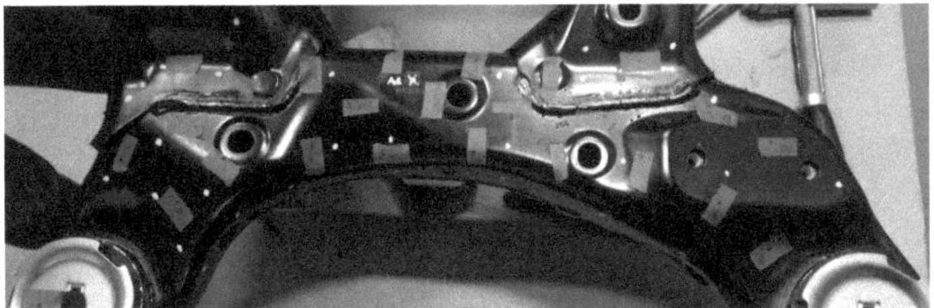

Figure 3: Grid points marked on the frame.

in Finite Element Methods. In our case, it helps us to better visualise the structure's surfaces (point clouds do not give a good plastic impression of the object's movements). A comprehensive listing of coordinate values and triangulation tables can be found in our Appendix. We decided to mount the accelerometers to the long and the short side to minimise the risk of positioning them on nodal lines. The positions are marked with circles in Fig. 2. Namely, our channel setup is:

- Channel 1: Force

- Channel 2: Accelerometer Point 12 (left in Fig. 2)

- Channel 3: Accelerometer Point 32 (centre in Fig. 2)

- Channel 4: Accelerometer Point 75 (right in Fig. 2)

We set up our laboratory equipment similar to the abstract sketch as seen in Figure 4. We excite our test object using hammer excitation. It is cost effective and provides easy handling considering the number of measurement points. A soft rubber tip provides enough excitation energy and avoids the risk of double hits. The signal of the force transducer and the three ICP accelerometers are transferred to a sufficient energy level by a simple amplifier. The signal is then passed to an A/D interface with included Fast Fourier Transform (FFT) analyser. The electronic system is managed by a proper MATLAB program called SIGLAB. A comprehensive list of lab components can be found in the Appendix.

We want to cover a required frequency range of 80 - 500 Hz. The upper frequency limit determines our sampling frequency f_s. To sample the signal adequately, we have to follow the **Shannon sampling theorem** [1], stating that the sampling frequency has to be at least twice the upper frequency limit f_u

$$f_{s,\text{opt.}} \geq 2\,f_u \tag{1}$$

and the built-in anti-aliasing filters will cut off signal components larger than f_u to avoid information mixing in the frequency domain of the FFT. As no filters can be ideal (infinitely steep cut off), the practical sampling rate is defined as

$$f_s = 2.56\,f_u. \tag{2}$$

We add an additional buffer and set $f_u = 1000$ Hz. The frequency resolution in the frequency domain should be at least $\Delta f = 1$ Hz. This will influence our time record length to

$$\Delta t = \frac{1}{\Delta f}. \tag{3}$$

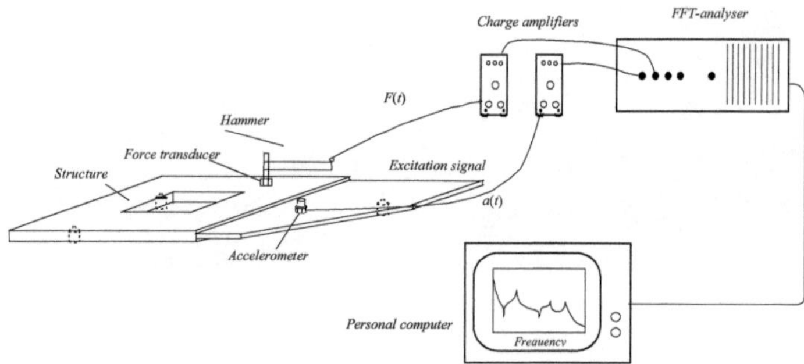

Figure 4: Schematised laboratory setup [1]

In the measuring software, we have to reverse engineer these equations, as we can only set the number of recorded samples per record. We use the maximum $N = 8192$ and in combination with the sampling frequency we get

$$\Delta f = \frac{f_s}{N} = 0.3125 \text{ Hz}. \tag{4}$$

$$\Delta t = \frac{1}{\Delta f} = 3.2 \text{ s} \tag{5}$$

As we like to make use of the noise-reduction in signals, we use the built-in averaging function. The measurement is automatically triggered with the pulse in the force transducer's signal. The input channels' sensitivity were tuned to fit to the different signal levels to obtain the best possible signal to noise ratio. A more detailed insight in the signal treatment will be given in the paragraphs hereafter.

3   Signal analysis

The signal processing is guided by the diagram which is depicted in Fig 5. First of all, the time signals of the impulse hammer's force and all acceleration signals are filtered in order to truncate frequencies which are larger than half of the sampling frequency. This procedure avoids aliasing, which would occur when digitalising the signals by using the A/D converter.

Measured signals in experiments are commonly non-periodic and have a finite signal length. For such signals, algorithms of the FFT would cause leakage errors, which lead to wrong frequency spectra. In order to reduce the influence of non-periodic and finite signals, weighting windows are normally used. The main characteristic of all windows used for FFT is that they start and end with zero. Since we use a hammer excitation, we assume our signals to be zero before the excitation and zero at the end of the measurement periods. Therefore, weighting windows were not used in our experiments. Next, the fast fourier transform is performed for the accelerations and excitation force. Based on the FFT, the averaged auto spectrum is computed for all signals, as well as the averaged cross spectrum between excitation force and all accelerations. Since we perform five measurements per excitation position, in order to lower the influence of noise and

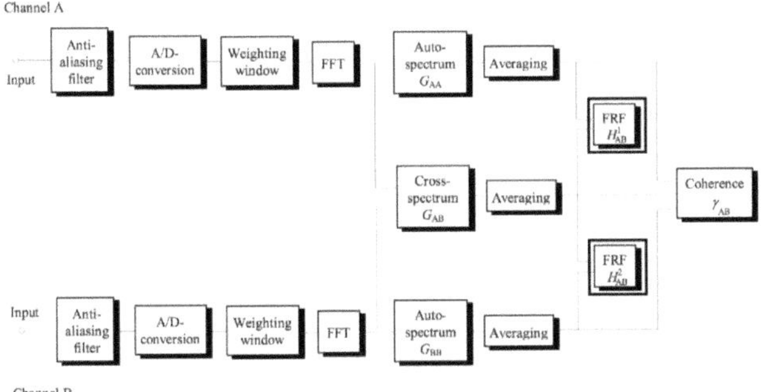

Figure 5: Signal processing in a FFT analyzer [1]

possible disturbances, the auto and cross spectra of all measurements are averaged. Then, the frequency response functions

$$h_{nm}(\omega) = \frac{\ddot{x}_n}{F_m} \tag{6}$$

between the acceleration $\ddot{x}$ at position n and force F at position m are estimated by dividing the cross spectra by the auto spectra of the excitation force. This approach is established since it provides the best FRF estimation. Finally, a calibration of all FRFs is performed. To this end, a known mass was excited with a hammer hit. The constant value for low frequencies of the calibration FRF as well as the mass are then used to calibrate the FRFs of the structure. Furthermore, the coherence function is determined. It is an indicator for linear relation between two signals.

Fig. 6 shows a small selection of various FRFs and the corresponding coherence in the frequency range of interest. One can see that some FRFs have common resonance peaks which indicates eigenfrequencies. Moreover, the coherence functions are close to one. That indicates linearity between input and output signal and good signal quality, respectively. However, the coherence drops if the corresponding FRF has an anti resonance. The measurement position 125 is located on the the gear while position 12 is on the top of the frame. It can clearly be seen, that the corresponding FRF $h_{12,125}$ has a low magnitude for higher frequencies. This means, that the impact of the hammer to position 12 is lower, if the forces are transferred through the soft bushings. A well isolating behaviour of the bushings can be concluded. However, the coherence within the high frequencies becomes worse as it shows more noise. Therefore, the signal should not be interpreted for very high frequencies, say above 600 Hz, which are not shown in the figure.

Another interesting fact is visible in Fig. 6. All FRFs have small disturbance which are noticeable as small local peaks. For instance, two of them occur within the frequency range from 150 Hz to 200 Hz. These disturbances emerge in the entire range and have a distance to each other of approx. 20 Hz. Because of this regularity, they cannot be caused by the investigated structure itself. The reason for the appearance must be in the in the setup of the measurement instruments. We assume that the disturbances are leakage errors, which are caused by omitting weighting windows.

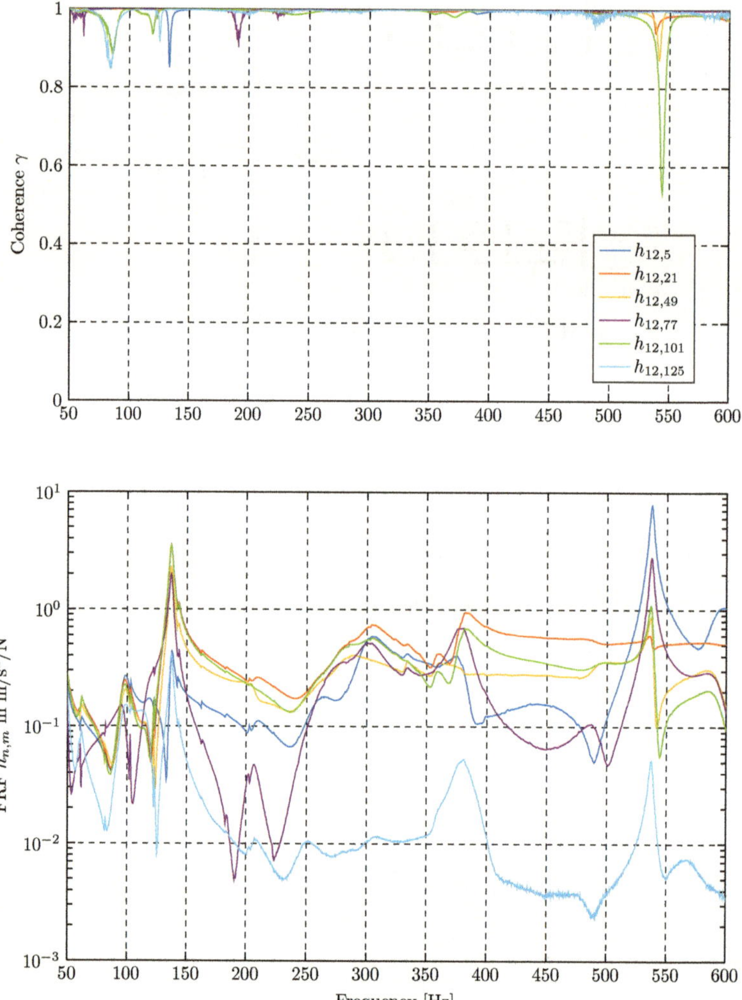

Figure 6: Typical FRFs and coherence functions

4   Mode extraction

Theory

The measured FRFs are used to determine all parameters of a model with a finite number of degrees of freedom. Such a model consists of a mass matrix $\mathbf{M}$ and stiffness matrix $\mathbf{K}$ and is described as follows:

$$\mathbf{M}\ddot{\mathbf{x}} + \mathbf{K}\mathbf{x} = \mathbf{F}. \tag{7}$$

The state vector $\mathbf{x}$ consists of all displacements, whereas $\mathbf{F}$ contains all external forces acting at the corresponding positions. The Laplace transform of the homogeneous formulation of Eqn. 7 with the Laplace variable s yields to the eigenvalue problem

$$\left(\mathbf{M}s^2 + \mathbf{K}\right)\mathbf{x} = 0, \tag{8}$$

which has the system poles s_n and the eigenvectors φ_n. By means of $\mathbf{\Phi}$, which contains the eigenvectors in its columns, the physical state vector can be defined as a linear combination of eigenvectors and modal states ξ:

$$\mathbf{x} = \mathbf{\Phi}\xi. \tag{9}$$

Using this definition, the modal formulation of Eqn. 7 with uncoupled system equations is obtained. Solving the n-th equation for the modal coordinate gives

$$\xi_n = \frac{1}{s^2\mu_n + \kappa_n}q_n \tag{10}$$

with the modal mass μ_n, modal stiffness κ_n and modal force q_n. Applying Eqn. 9 yields to

$$x_k = \sum_n \frac{\varphi_{kn}\varphi_{mn}}{s^2\mu_n + \kappa_n}F_m. \tag{11}$$

The FRFs can now be obtained by rearranging and introducing the system poles $s_n^2 = \frac{\kappa_n}{\mu_n}$:

$$h_{km} = \frac{x_k}{F_m} = \sum_n \frac{\phi_{kn}\phi_{mn}}{s_n^2 + s^2}. \tag{12}$$

This equation shows that the eigenvalues and eigenvectors are linked to the FRF. Therefore, the mode parameters of the model can be calculated using the model's FRFs h_{km}. Since we use the complex exponential method, partial fraction expansion is applied to the right hand side of Eqn. 12,

$$h_{km} = \sum_n \frac{{}_nR_{km}}{s - s_n} + \left(\frac{{}_nR_{km}}{s - s_n}\right)^*, \tag{13}$$

which now uses the residues ${}_nR_{km}$. Inverse Laplace transform gives the impulse response function (IRF)

$$j_{km}(t) = \sum_n {}_nR_{km}e^{s_n t} + {}_nR_{km}^* e^{s_n^* t}. \tag{14}$$

The IRF is needed for a numerical fit, which tries to match the measured FRF as effectively as possible. This approach, which operates in the z-domain, is known as Prony's method and it is commonly used in digital filter synthesis. We define a residual r_{km} which is the error between the z-transform of the measured FRF $\tilde{H}_{km}(z)$ and the approximation $H_{km}(z)$

$$r_{km} = H_{km}(z) - \tilde{H}_{km}(z). \tag{15}$$

The approximation is modelled as a ratio between two polynomials

$$H_{km}(z) = \frac{b_0 + \cdots + b_q \cdot z^{-q}}{1 + \cdots + a_p \cdot z^{-p}}. \tag{16}$$

Using the definition of the approximation and the relation

$$\tilde{H}_{km}(z) = \sum_{n=0}^{\infty} z^{-n} \tilde{j}_{km} \tag{17}$$

between z-transformed FRF and IRF, a linear optimisation problem arises:

$$r_{km} \cdot \left(1 + \cdots + a_p \cdot z^{-p}\right) = \sum_{n=0}^{\infty} z^{-n} \left(b_n - \left(\tilde{j}(n)\right) + \cdots + a_p \tilde{j}(n-p)\right) \tag{18}$$

Finally, the residual is set to zero so that the left hand side vanishes. The right hand side satisfies the equation if the coefficients for powers in z^{-n} itself vanish. Therefore, theoretically infinite conditions are found to determine the coefficients a_n and b_n of denominator and numerator. We use the Shank's approximation, which is characterised by the use of an overestimated system of equations. That makes the method more resistant against noise and other disturbances. After the coefficients for the approximated FRF $H_{km}(z)$ are found, the eigenfrequencies ω_n are computed from the poles p_n according to

$$\omega_n = -i\omega_s \ln(p_n) \tag{19}$$

with the sampling frequency ω_s. The modal damping is computed by

$$\eta_n = 2 \cdot \frac{\text{Im}(\omega_n)}{\text{Re}(\omega_n)}. \tag{20}$$

The residues $_nR_{km}$ and eigenvectors, respectively, are determined by expanding the approximated FRF in a rational fraction series, compare Eqn. 13.

Application-oriented advises

At the end of this section, some practical advises on the mode extraction procedure are given. Fig. 7 depicts a pole stability diagram. After having defined the maximum polynomial order of the pole estimation, the algorithm will perform estimations from polynomial order one to the given number. Markers with a higher y-position represent estimations with a higher polynomial order. The colouring of the stars depends on an error estimation of the poles' variation in terms of angle and absolute value. Green markers represent fits that generally remain below the error threshold and are considered as stable. It is now the art to vary the polynomial order to obtain stable (=green) poles but at the same time not capturing noise in the curve fit (e.g. in Fig. 8 at 124 Hz and 145 Hz). The mode indicator function (blue line in Fig. 7) helps us to determine if an estimated pole matches a real resonance peak. In the presented plot, one should not at all choose the most right pole, as it is obviously not a physical resonance. However, Fig. 8 proves that the FRF curve fit successfully removes the influence of signal noise once a stable poble has been chosen.

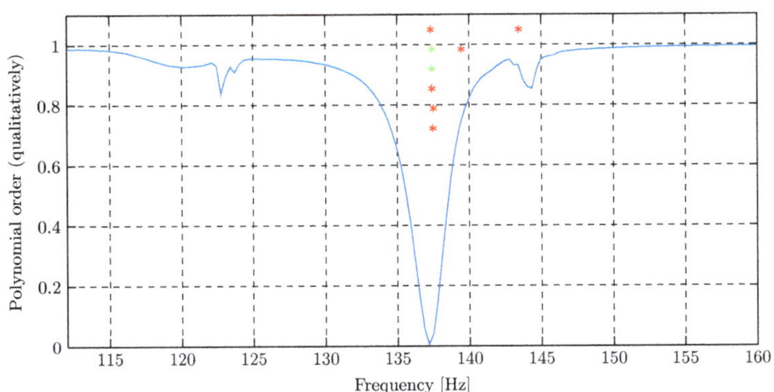

Figure 7: Typical FRFs and coherence functions

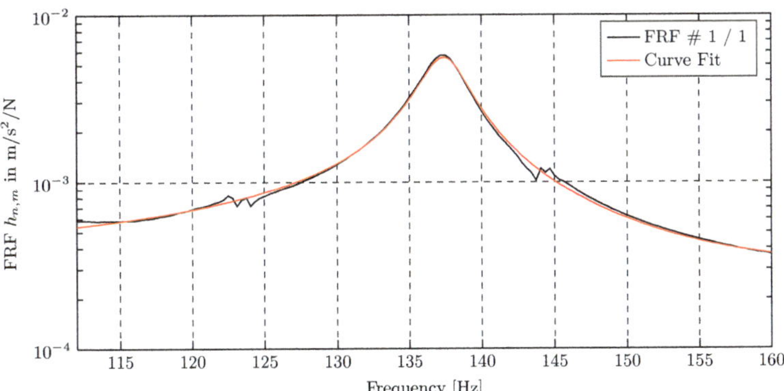

Figure 8: Typical curve fit of a resonance peak

5   Results

A straight-forward way to scan the measured FRFs for resonance peaks is the simple mode indicator function

$$I_m(\omega) = \sum_n |h_{nm}(\omega)|^2, \tag{21}$$

which is plotted in Fig. 9 for all three measurement channels. The numbering of the modes is consistent with the numbering in Tab. 1. On the basis of this plot, we decided which poles could be of interest in the given frequency range. Unfortunately, it is not always possible to capture all resonances with three recording channels, as the accelerometer could be placed on a mode's nodal line. We see such a case for example at approximately 250 Hz. Only channel three captures a resonance peak but the individual FRFs showed not enough quality to handle a curve fit adequately. Another problem may be introduced by modes which are located too close to each other, see channel four at 350 to 380 Hz. In this case, we decided to curve fit the FRF of channel three and assuming the risk of incorporating the influence of a smaller mode at 350 Hz. The resonance frequencies and modal damping values have been averaged for all three channels and results only were excluded if they could not capture a resonance properly.

Figures 10 and 11 depict screenshots of all investigated modes at their maximum deflection. To preserve comparability, all plots feature the same viewing angle. Attached to this report, find the corresponding AVI movies recorded in the most significant angles. As we wanted to give the client the opportunity to investigate the different modes interactively, a MATLAB GUI has been designed from scratch. At start up, the user is asked to name a MAT file containing the extracted modes. Point at `Modes1-10.mat`. Now, the animation starts and the frequency may be changed at the top left. Most important modal parameters are displayed here as well. The rotate tool of MATLAB Figures can be used to define an individual viewing angle. Speed and amplitude changes may help to better identify the mode shape characteristics. An included video function helps to record the animation and export it to an AVI movie.

It was decided in the beginning to only perform measurements in z direction as laboratory time is limited and 134 DOFs already charges the former to capacity. However, this bears the risk of capturing modes with deflections in the x / y plane. To exclude these modes, we analysed a three-dimensional 25 DOF data set for out-of-plane modes. Mode shape candidates two and six have to be excluded due to the above-mentioned reasons. Mode number ten can not be verified as the backup data set does not provide resonance peaks at this frequency.

Modes one and three feature a - what we call - two mass oscillator behaviour. The assembled components remain undeformed but perform an up and down movement with 180 degrees phase shift. The soft bushings seem to allow a relatively large tilting of one side of the gear box, respectively. These two modes could have most influence on the bushings life period, if the excitation energy level is high enough in the lower frequency range during normal driving. For all resonance frequencies higher than 100 Hz, we cannot distinguish this intensive gear frame interaction anymore. It can be expected that starting from this "isolation frequency", the bushings isolate well the frame modes from the gear box, which allows an unhindered drive train process. The residual modes 4, 5, 7, 8 and 9 consequently show well isolated torsion and bending modes that are even comparable with the modes of a simple plate. Now, we can benefit from our second layer of measurement points. Figure 12 clearly depicts the distinct characteristics of each mode shape. Most noticable may be mode number seven. The long edges perform an antiphase bending and consequently effect a torsion in the short ones. On top of that, the elastomer inlays of the end connectors introduce phase shifts of 180 degrees (cf. Fig 11a).

The modal damping factors in Tab. 1 generally give an insight in how easy the distinct modes may get excited by a given input signal. From this point of perspective, the tilting modes (no.

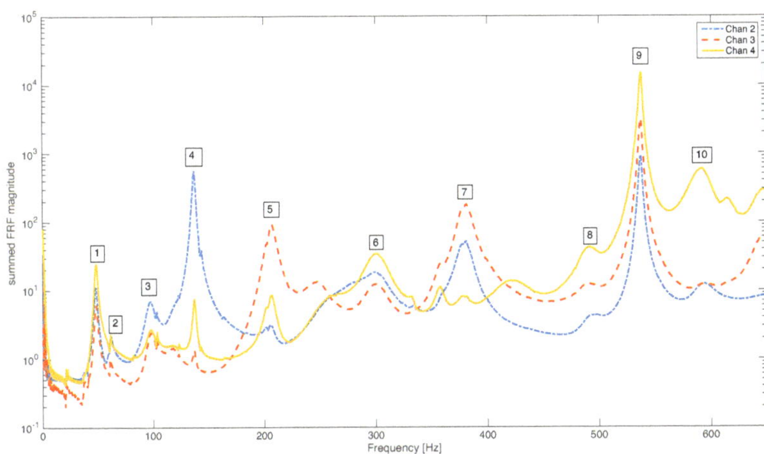

Figure 9: Summed FRF magnitudes.

Table 1: Extracted Modes

Mode	Channel	Modal Damping [%]	Freq. [Hz]	Name
1	4	6.8	48.4	two mass oscillator (tilt right)
2	2	4.0	60.5	- y dominated movement -
3	2	8.8	97.2	two mass oscillator (tilt left)
4	2	2.0	137.2	torsion 1st
5	3	3.6	206.2	bending 1st
6	4	7.4	298.5	- x dominated movement -
7	3	5.0	380.2	bending antiphase
8	4	0.6	489.9	bending 2nd
9	4	0.6	537.5	bending short side
10	4	4.0	591.2	- similar to 9, not verified -

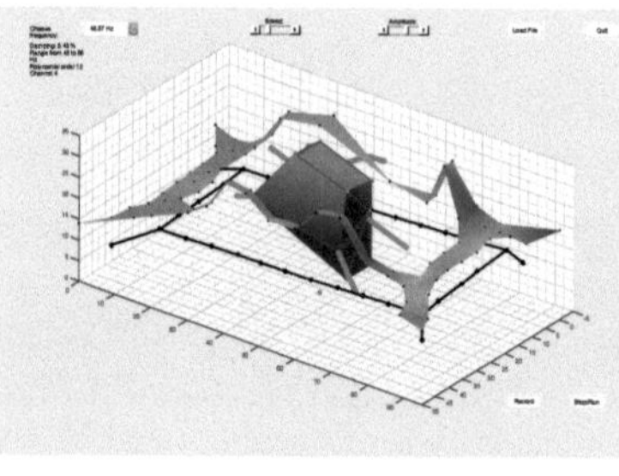

(a) Mode 1

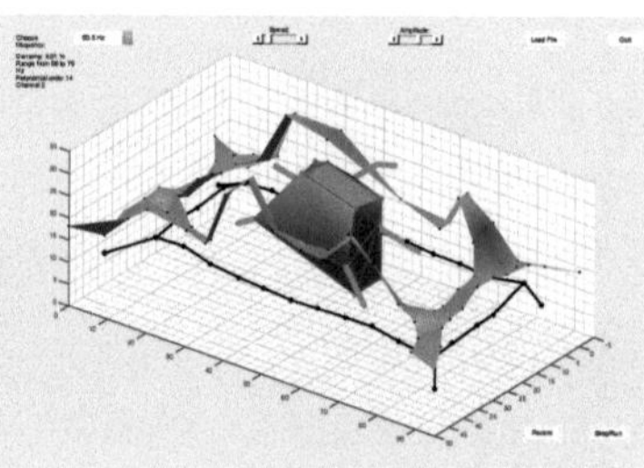

(b) Mode 2

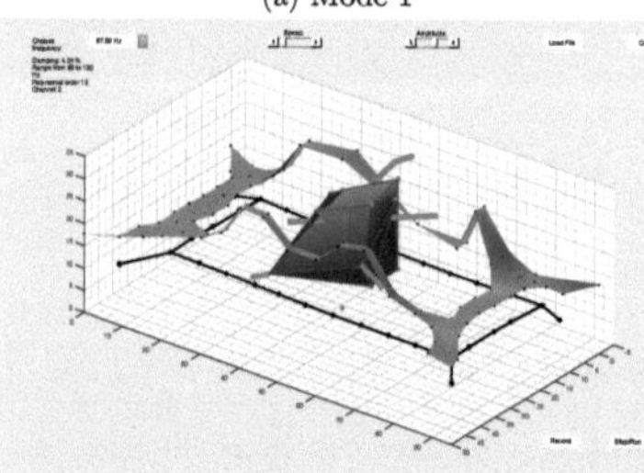

(c) Mode 3

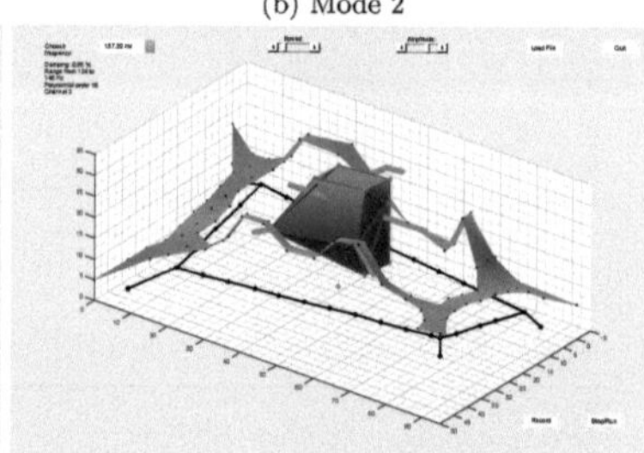

(d) Mode 4

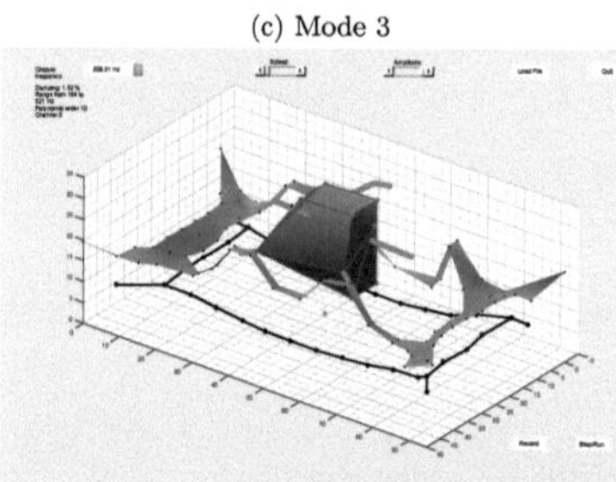

(e) Mode 5

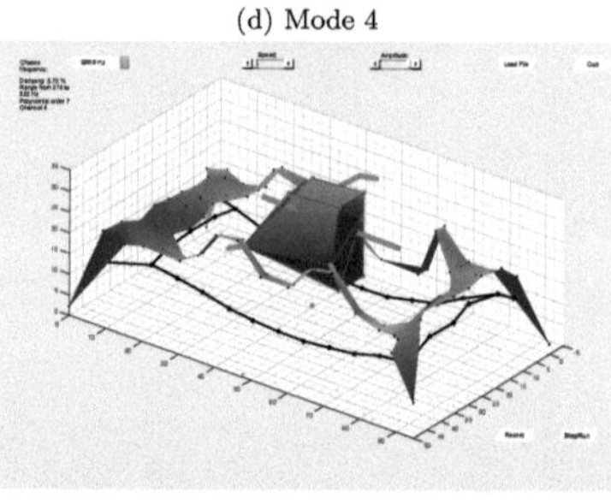

(f) Mode 6

Figure 10: Mode shapes 1 - 6

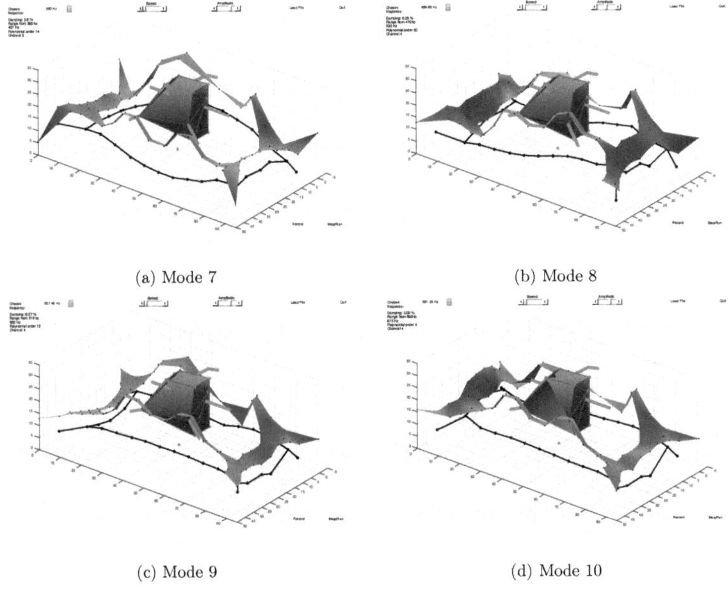

(a) Mode 7

(b) Mode 8

(c) Mode 9

(d) Mode 10

Figure 11: Mode shapes 7 - 10

one and three) are well damped. This can be explained by high energy dissipation of the soft bushings, which are strongly deformed when the structure oscillates in these modes. Anyhow, they can of course be excited if the energy level is sufficiently high. On the other side, the first torsion and the first and second bending mode are the most undamped modes in the frequency range of interest.

Part of a critical discussion of our findings should always be the advice to compare the obtained damping factors and resonance frequencies with the results of other groups to obtain a whole picture on the problem, but we only found minor variations for the modes that we have verified with our second data set. For future investigations, it would be advisable to perform more comprehensive studies in the x / y directions. Furthermore, it would be interesting to determine the mode shapes for other boundary conditions. The free suspension of the frame may be too artificial compared to the mounted setup of the frame in a car (with more or less fixed end connectors).

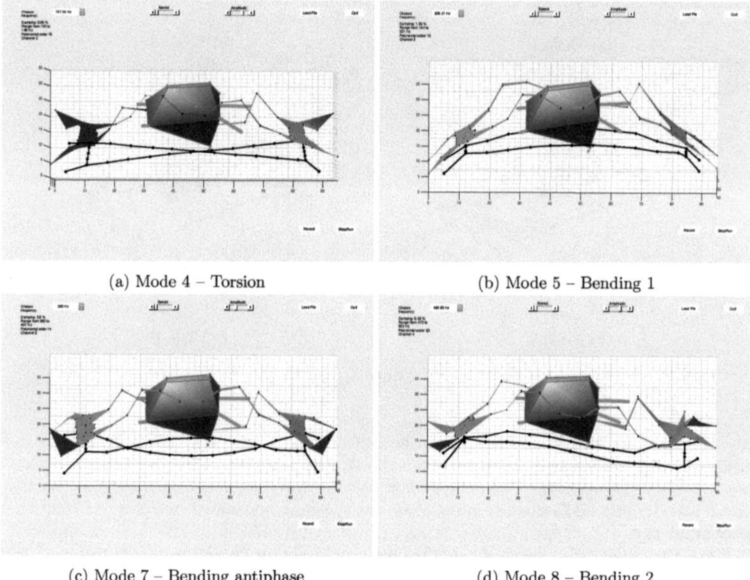

(a) Mode 4 – Torsion

(b) Mode 5 – Bending 1

(c) Mode 7 – Bending antiphase

(d) Mode 8 – Bending 2

Figure 12: Side views of various mode shapes

6   Appendix

Laboratory setup

Impact hammer: Dytran, Dynapulse 5800A3, serial number 469
Accelerometers: Brüel & Kjær 4507 B005, serial numbers 10062,10063 & 10064
Data acquisition system: DSP Technology, Siglab 20-42, Serial number 11315

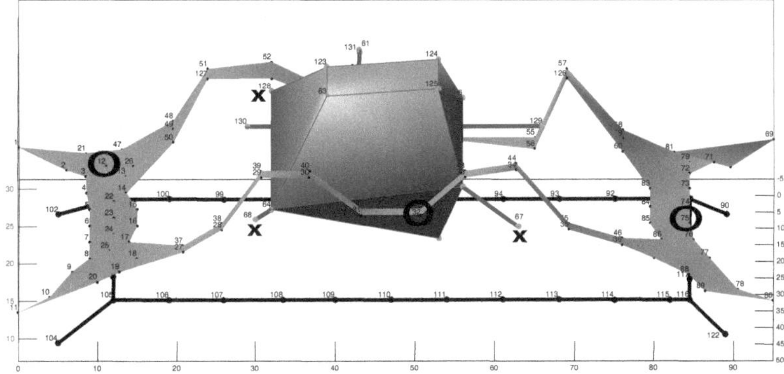

Figure 13: Measurement grid

Table 2: Measurement grid (coordinates in centimeter)

Node	x	y	z	Node	x	y	z	Node	x	y	z
1	0	0	13.5	46	35.5	76	17	91	5.5	84.5	9
2	8	6	14	47	4	13	15	92	5.5	75	9
3	11	8.5	14.5	48	4.50	19.5	19	93	5.50	68	9
4	16	8.60	14.5	49	7	19.5	19	94	5.5	61	9
5	21	9	14.5	50	11	19.5	19	95	5.5	54	9
6	26	9	14.5	51	4.5	24	26	96	5.5	47	9
7	31	9	14.5	52	-1	32	24.5	97	5.5	40	9
8	36	9	14.5	53	4	32	24.5	98	5.5	33.5	9
9	40	6.8	14.5	54	5	52	17	99	5.5	26	9
10	43	4	12.5	55	5	65	17	100	5.5	19	9
11	50	0	13.5	56	8.5	65	17	101	5.5	12	9
12	8	11	14.5	57	4.5	69	26	102	5.5	5	7
13	11	13.5	14.5	58	5	76	18	103	22	12	9
14	16	13.6	14.5	59	7.5	76	18	104	44.5	5	7
15	21	15	14.5	60	11.5	76	18	105	36	12	9
16	26	15	14.5	61	1	43	27	106	36	19	9
17	31	14	14.5	62	9	43	27.5	107	36	26	9
18	36	15	14.5	63	22	39	30	108	36	33.5	9
19	40	12.8	14.5	64	29	32	18	109	36	40	9
20	43	10	14.5	65	18	56	28	110	36	47	9
21	4	8.5	14.5	66	29	56	21	111	36	54	9
22	18.5	12	14.5	67	32	63	17	112	36	61	9
23	23.5	12	14.5	68	32	30	18	113	36	68	9
24	28.5	12	14.5	69	0	95	14.5	114	36	75	9
25	33.5	11.5	14.5	70	5	89.5	13	115	36	82	9
26	8	14.5	14.5	71	8	87.5	15	116	36	84.5	9
27	37.5	20.8	16	72	11.5	84.5	15	117	30	84.5	9
28	37.5	25.8	19	73	16	84.5	15	118	21.5	84.5	9
29	37.5	30.8	26	74	21.5	84.5	15	119	16	84.5	9
30	37.5	36.8	26	75	26.5	84.5	15	120	13	12	9
31	37.5	43.3	21	76	31.5	85	15	121	29	12	9
32	37.5	50.8	21	77	37	87	15	122	42	89	7
33	37.5	56.3	26	78	42	90.5	13	123	13	39	30
34	37.5	62.8	27	79	8	84.5	15	124	13	53	31
35	37.5	69.3	19	80	50	95	15	125	22	53	31
36	37.5	76	17	81	5	82.5	15	126	7.5	69	26
37	35.5	20.8	16	82	8	80.5	15	127	7.5	24	26
38	35.5	25.8	19	83	16	79.5	15	128	8	32	24.5
39	35.5	30.8	26	84	21.5	79.5	15	129	17.5	66	24
40	35.5	36.8	26	85	26.5	79.5	15	130	17.5	29	24
41	35.5	43.3	21	86	31.5	81	15	131	-5	43	24
42	35.5	50.8	21	87	37	80	15	132	22	53	11
43	35.5	56.3	26	88	42	84.5	15	133	13	53	11
44	35.5	62.8	27	89	47	86.5	15	134	31	52.5	11
45	35.5	69.3	19	90	5.5	89	7				

Table 3: Triangulation

# element	node 1	node 2	node 3	# element	node 1	node 2	node 3
1	11	20	10	48	81	71	70
2	20	9	10	49	69	81	70
3	20	19	9	50	82	72	79
4	19	8	9	51	79	72	71
5	19	18	8	52	83	72	60
6	18	25	8	53	82	60	72
7	8	25	7	54	72	83	73
8	18	17	25	55	84	74	73
9	25	17	7	56	83	84	73
10	17	24	7	57	84	85	74
11	17	16	24	58	85	75	74
12	7	24	6	59	85	86	75
13	24	23	6	60	75	86	76
14	24	16	23	61	86	87	76
15	23	22	5	62	87	77	76
16	23	15	22	63	87	88	77
17	22	15	14	64	88	78	77
18	22	14	4	65	88	89	78
19	5	22	4	66	89	80	78
20	14	3	4	67	36	87	86
21	14	13	3	68	46	36	86
22	13	12	3	69	46	35	36
23	12	2	3	70	46	45	35
24	12	21	2	71	45	34	35
25	21	1	2	72	45	44	34
26	13	26	12	73	44	33	34
27	26	47	12	74	44	43	33
28	14	50	13	75	43	42	32
29	13	49	26	76	43	32	33
30	13	50	49	77	42	41	31
31	26	49	48	78	42	31	32
32	26	48	47	79	41	40	30
33	50	127	49	80	41	30	31
34	127	51	49	81	40	39	29
35	49	51	48	82	40	29	30
36	52	51	53	83	39	38	28
37	53	51	127	84	39	28	29
38	52	53	54	85	38	37	27
39	55	54	56	86	38	27	28
40	55	56	57	87	37	18	27
41	58	126	59	88	17	18	37
42	59	126	60	89	5	6	23
43	58	59	81	90	23	16	15
44	81	59	82	91	21	12	47
45	82	59	60	92	18	19	27
46	82	79	81	93	56	126	57
47	81	79	71	94	57	126	58

MATLAB Codes

```matlab
function [XFER,COH,dF,freq]=importData(chan)
close all
[FileName,PathName] = uigetfile('*.mat','Specify FRF file');
n=str2num(FileName(end-6:end-4));
XFER=NaN(3201,n);
COH=NaN(3201,n);
scrsz = get(groot,'ScreenSize');
figure('Position',[1 1 scrsz(3) scrsz(4)])

for jj=1:n
    FileIn = [FileName(1:end-7),num2str(jj,'%01.3d')];
    load([PathName,FileIn]);

time=vna_data.tdxvec;
y(:,1)=vna_data.scmeas(1).tdmeas;
y(:,2)=vna_data.scmeas(2).tdmeas;
y(:,3)=vna_data.scmeas(3).tdmeas;
y(:,4)=vna_data.scmeas(4).tdmeas;
aspec(:,1)=vna_data.scmeas(1).aspec;
aspec(:,2)=vna_data.scmeas(2).aspec;
aspec(:,3)=vna_data.scmeas(3).aspec;
aspec(:,4)=vna_data.scmeas(4).aspec;

dF=vna_data.rbw;
freq=vna_data.fdxvec;
xfer(:,2)=vna_data.xcmeas(1, 2).xfer;
coh(:,2)=vna_data.xcmeas(1, 2).coh;
xfer(:,3)=vna_data.xcmeas(1, 3).xfer;
coh(:,3)=vna_data.xcmeas(1, 3).coh;
xfer(:,4)=vna_data.xcmeas(1, 4).xfer;
coh(:,4)=vna_data.xcmeas(1, 4).coh;

XFER(:,jj)=xfer(:,chan);
COH(:,jj)=coh(:,chan);

%% Plotting
figure('Position',[1 1 scrsz(3)/2 scrsz(4)])
subplot(2,1,1)
plot(time,y_f)
title('Time Domain')
subplot(2,1,2)
plot(freq,aspec_f)
title('Autospectra')

figure('Position',[scrsz(3)/2 1 scrsz(3)/2 scrsz(4)])
subplot(2,1,1)
plot(time,y_acc)
title('Time Domain')
subplot(2,1,2)
plot(freq,aspec_acc)
title('Autospectra')

subplot(2,1,1)
plot(freq,coh)
title('Coherence')
grid minor
hold on
subplot(2,1,2)
```

```matlab
59   semilogy(freq,abs(xfer))
60   title('Transfer Function')
61   grid minor
62   hold on;
63
64   end
```

Algorithm 1: Data import from SIGLAB files

```matlab
1    function [MODE,RESDAMP]=analyseData(fpos,file,prefix)
2    addpath('ComplEst');
3    close all                    % Necessary for correct figures!
4    nres=size(fpos,1);           % No. of Frequency ranges
5    % MODE=NaN(size(XFER,2),nres);
6    for ii=1:nres
7    disp(['Analysing from ' num2str(fpos(ii,1)) ' Hz to ' num2str(fpos(ii,2)) ' Hz -
         PolyOrder ' num2str(fpos(ii,3)) ' on Channel ' num2str(fpos(ii,4))])
8    % Load Data
9    load([file num2str(fpos(ii,4)) '.mat'])
10   % Init
11   Spec_size = length(XFER);    % Number of points in spectrum incl 0 Hz.
12   Xaxis = (0:Spec_size-1)*dF;  % Frequency axis.
13
14   disp(['Recommended Frequency: ' num2str(fpos(ii,5))])
15
16   % Truncate elements in FRF matrix to selected frequency range
17   [At,ft] = frftrunc(XFER,Xaxis,fpos(ii,1),fpos(ii,2));
18
19   % Convert from accelerance matrix to mobility matrix
20   H = cvfrfa2v(At,ft);
21
22   % Calculate Mode Indicator Function (MIF)
23   mif = modeind1(H);
24
25   % Calculate impulse response matrix
26   [h,t,fs] = impresp(H,ft);
27
28   % Estimate poles with complex exponentials method. Nr of poles = 60.
29   poles = complexp(h,fs,min(300,length(ft)),fpos(ii,3),mif,ft,fpos(ii,1),fpos(ii,2));
30
31   % Calculate eigen-frequency and modal damping from poles
32   [resfreq,zeta] = poles2fd(poles);
33
34   % Estimate residues from poles
35   [residues,residuals] = pol2resf(H,ft,poles,fpos(ii,1),fpos(ii,2),0.1);
36
37   % Calculate mode shapes and modal participation factors from residues
38   [L,moder] = res2mpf(residues);
39
40   % Save Data
41   close all
42   if size(moder,1)>1
43       MODE=moder';
44       RESDAMP(1,:)=resfreq';
45       RESDAMP(2,:)=zeta';
46   else
47   MODE(:,ii)=moder';
48   RESDAMP(1,ii)=resfreq;
49   RESDAMP(2,ii)=zeta;
50   end
51   RESDAMP(3:6,ii)=fpos(ii,1:4)';
```

```
52  end
53  currentFolder = pwd;
54  save([currentFolder,'/ModesUlf/',prefix],'MODE','RESDAMP');
```

Algorithm 2: Data analysis

```
1   function animateData(action)
2   % This function allows to animante all eigenvectors. Point to file
3   % Modes1-10.mat. The file mesh.mat must be available in the active
4   % directory.
5
6   persistent FIG. H. T. L1. L2. P1. P2. Label. gear. bodyfloor. bodytop. face3. ...
7       face4. MODE RESDAMP NumMode. coordinate. element. elementbase. Xline. Yline. ...
8       elementgear. normeigenvec. RUN. Popup. V. Slider. AmpMax. Xlinegear. ...
9       Ylinegear. Button. REC. writerObj.
10
11  %% First call?
12  if nargin==0;
13      action='initialize';
14  end
15
16  %% Initializes the program
17  if strcmp(action,'initialize')
18      % sets settings for the first mode
19      NumMode.=1;
20      V.=0.1;
21      AmpMax.=5;
22      REC.=0;
23      % Calls other functions
24      animateData('LoadData')
25      animateData('LoadMeshData')
26      animateData('CreateFirstPlot')
27      animateData('CreateGUI')
28      animateData('animate')
29  end
30
31  %% load files
32  if strcmp(action,'LoadData')
33      RUN.=0;
34      currentFolder=pwd;
35      [FileName, PathName]=uigetfile('*.mat','Select the .mat file',[currentFolder '/
            modes']);   % loads MODE and RESDAMP
36      load([PathName FileName]);
37      NumMode.=1;
38      normeigenvec. = AmpMax.*MODE(:,NumMode.)/max(abs(MODE(:,NumMode.)));% normalized
            eigenvector with 5 cm as maximal value
39      RUN.=1;
40  end
41
42  %% Loading and renaming of mesh data
43  if strcmp(action,'LoadMeshData')
44      load('mesh.mat')
45      coordinate.=coordinate;
46      bodyfloor.=bodyfloor;
47      bodytop.=bodytop;
48      gear.=gear;
49      element.=element;
50      elementbase.=elementbase;
51      elementgear.=elementgear;
52      face3.=face3;
53      face4.=face4;
```

```matlab
54      clear coordinate bodyfloor bodytop gear element elementbase face3 face4
55  end
56
57  %% Creates the first figure
58  if strcmp(action,'CreateFirstPlot')
59
60      scrsz = get(groot,'ScreenSize');
61      FIG_=figure('Position',[1 1 scrsz(3) scrsz(4)]);
62
63
64      % coordinates of different parts
65      xtop=coordinate_(bodytop_,2);
66      ytop=coordinate_(bodytop_,3);
67      ztop=coordinate_(bodytop_,4);
68
69      xfloor=coordinate_(bodyfloor_,2);
70      yfloor=coordinate_(bodyfloor_,3);
71      zfloor=coordinate_(bodyfloor_,4);
72
73      xgear_=coordinate_(gear_,2);
74      ygear_=coordinate_(gear_,3);
75      zgear_=coordinate_(gear_,4);
76
77      % Plot all measured points
78      H_(1)=scatter3(xtop,ytop,ztop,10,'black','filled');
79      hold on;
80      grid minor
81      axis equal
82      zlim([0 35])
83      daspect([1 1 1]);
84      H_(2)=scatter3(xfloor,yfloor,zfloor,60,'black','filled');
85      H_(3)=scatter3(xgear_,ygear_,zgear_,60,'red','filled');
86
87      % Surface
88      T_(1)=trisurf(element_(:,1:3),coordinate_(:,2),coordinate_(:,3),coordinate_(:,4)
89          ,'FaceColor',[0.0641    0.5570    0.8240],'EdgeAlpha',0);
90      % Plot line model
91      Xline_=[coordinate_(elementbase_(:,1),2) coordinate_(elementbase_(:,2),2)]';
92      Yline_=[coordinate_(elementbase_(:,1),3) coordinate_(elementbase_(:,2),3)]';
93      Zline=[coordinate_(elementbase_(:,1),4) coordinate_(elementbase_(:,2),4)]';
94      L1_=plot3(Xline_,Yline_,Zline);
95      set(L1_,'Color','black','LineWidth',4)
96
97      Xlinegear_=[coordinate_(elementgear_(:,1),2) coordinate_(elementgear_(:,2),2)]';
98      Ylinegear_=[coordinate_(elementgear_(:,1),3) coordinate_(elementgear_(:,2),3)]';
99      Zlinegear=[coordinate_(elementgear_(:,1),4) coordinate_(elementgear_(:,2),4)]';
100     L2_=plot3(Xlinegear_,Ylinegear_,Zlinegear);
101     set(L2_,'Color',[0.5 0 0],'LineWidth',8)
102
103     % Plot Faces
104     P1_=patch('Vertices',coordinate_(:,2:4),'Faces',face4_,...
105         'FaceVertexCData',[1 0 0],'FaceColor','flat','FaceAlpha',1,'EdgeColor',[0.5
106             0 0]);
107     P2_=patch('Vertices',coordinate_(:,2:4),'Faces',face3_,...
108         'FaceVertexCData',[1 0 0],'FaceColor','flat','FaceAlpha',1,'EdgeColor',[0.5
109             0 0]);
110     % Viewing settings
111     view(127,27);
        lightangle(127,127)
```

```matlab
112       set(T_,'FaceLighting' , 'gouraud','AmbientStrength',0.3,'DiffuseStrength',0.8,'
              SpecularStrength',0.3,'SpecularExponent',25);
113   end
114
115   %% Creats all UI buttons and sliders
116   if strcmp(action,'CreateGUI')
117       Popup_ = uicontrol('Style', 'popup',...
118           'String', strcat(cellstr(num2str((floor(100*RESDAMP(1,:))/100)'))', ' Hz'),
              ...
119           'Callback', 'animateData(''changeMode'')',...
120           'Units','Normalized','Position',[0.13 0.92 0.1 0.045]);
121
122       Label_(1)=uicontrol('style','text','String','Choose frequency:',...
123           'Units','Normalized','Position',[0.05 0.92 0.08 0.04],...
124           'HorizontalAlignment', 'left');
125
126       Label_(2)=uicontrol('style','text','String',...
127           {['Damping: ' num2str(2*round(100*RESDAMP(2,NumMode_))/100) ' %'],...
128           ['Range from ' num2str(RESDAMP(3,NumMode_)) ' to ' num2str(RESDAMP(4,
                  NumMode_)) ' Hz'],...
129           ['Polynomial order ' num2str(RESDAMP(5,NumMode_))],...
130           ['Channel ' num2str(RESDAMP(6,NumMode_))]},...
131           'Units','Normalized','Position',[0.05 0.82 0.1 0.1],...
132           'HorizontalAlignment', 'left');
133
134       Label_(3)=uicontrol('style','text','String','Speed:','Units','Normalized',...
135           'Position',[0.4 0.95 0.08 0.03]);
136
137       Label_(4)=uicontrol('style','text','String','Amplitude:','Units','Normalized',
              ...
138           'Position',[0.6 0.95 0.08 0.03]);
139
140       Slider_(1)=uicontrol('style','slider', 'Units','Normalized','Position',[0.4 0.93
              0.08 0.03],...
141           'String','Length 1', 'Min',0.05,'Max',0.2,'Value',V_,'Callback','animateData
              (''updateslider'')');
142
143       Slider_(2)=uicontrol('style','slider', 'Units','Normalized','Position',[0.6 0.93
              0.08 0.03],...
144           'String','Length 1', 'Min',1,'Max',15,'Value',AmpMax_,'Callback','
              animateData(''updateslider'')');
145
146       Button_(1)=uicontrol('style','pushbutton', 'Units','Normalized','Position',[0.8
              0.93 0.06 0.04],...
147           'String','Load File','Callback','animateData(''LoadData'');animateData(''
              CreateGUI'')');
148
149       Button_(2)=uicontrol('style','pushbutton', 'Units','Normalized','Position',[0.92
              0.93 0.06 0.04],...
150           'String','Quit','Callback','animateData(''stop'')');
151
152       Button_(3)=uicontrol('style','togglebutton', 'Units','Normalized','Position',[0
              .8 0.1 0.06 0.04],...
153           'String',{'Record'},'Callback','animateData(''record'')');
154
155       Button_(4)=uicontrol('Style','PushButton','Units','Normalized','Position',[0.9 0
              .1 0.06 0.04],...
156                       'Callback','animateData(''RunOrStop'')','String','Stop/Run');
157
158
159   end
160
```

```matlab
%% animation loop
if strcmp(action,'animate')
    RUN_=1;
    omegat=0;
    while ((RUN_==1) && (exist('FIG_','var')~=0));

        % New z coordinate_s
        zcoord=coordinate_(:,4)+real(normeigenvec_*exp(1i*omegat));
        zcoord([bodyfloor_ 132 133 134])=zcoord([bodyfloor_ 132 133 134])-2*real(
            normeigenvec_([bodyfloor_ 132 133 134])*exp(1i*omegat));

        % Update measuring points
        set(H_(1), 'ZData',zcoord(bodytop_));
        set(H_(2), 'ZData',zcoord(bodyfloor_));
        set(H_(3), 'ZData',zcoord(gear_));

        % Update surface
        delete(T_)
        T_(1)=trisurf(element_(:,1:3),coordinate_(:,2),coordinate_(:,3),zcoord,'
            FaceColor',[0.0641    0.5570    0.8240],'EdgeAlpha',0);
        set(T_,'FaceLighting' , 'gouraud','AmbientStrength',0.3,'DiffuseStrength',0
            .8,'SpecularStrength',0.3,'SpecularExponent',25);

        % Update Patches
        delete(P1_)
        P1_(1)=patch('Vertices',[coordinate_(:,2:3),zcoord],'Faces',face4_,...
            'FaceVertexCData',[1 0 0],'FaceColor','flat','FaceAlpha',1,'EdgeColor',[0.5
            0 0],'FaceLighting' , 'gouraud');
        delete(P2_)
        P2_(1)=patch('Vertices',[coordinate_(:,2:3),zcoord],'Faces',face3_,...
            'FaceVertexCData',[1 0 0],'FaceColor','flat','FaceAlpha',1,'EdgeColor',[0.5
            0 0],'FaceLighting' , 'gouraud');

        % Update lines
        Zline=(zcoord(elementbase_))';
        delete(L1_)
        L1_=plot3(Xline_,Yline_,Zline);
        set(L1_,'Color','black','LineWidth',3)

        Zlinegear=(zcoord(elementgear_))';
        delete(L2_)
        L2_=plot3(Xlinegear_,Ylinegear_,Zlinegear);
        set(L2_,'Color','red','LineWidth',8)

        % Section for recording videos
        if REC_==0;
            drawnow limitrate
        elseif REC_>0;
            drawnow
            if REC_==0.5
                writerObj_ = VideoWriter('Video.mp4', 'MPEG-4');
                open(writerObj_);
                REC_=1;
            end
            frame= getframe(FIG_);
            writeVideo(writerObj_,frame);
        elseif REC_==-1;
                close(writerObj_);
                REC_=0;
                drawnow limitrate
        end
        omegat=omegat+V_;
```

```matlab
218        end
219    end
220
221    %% Changes settings for animating another mode
222    if strcmp(action,'changeMode')
223        NumMode_=get(Popup_,'Value');
224        normeigenvec_ = AmpMax_*MODE(:,NumMode_)/max(abs(MODE(:,NumMode_)));
225        set(Label_(2), 'String',...
226            {['Damping: ' num2str(2*round(100*RESDAMP(2,NumMode_))/100) ' %'],...
227            ['Range from ' num2str(RESDAMP(3,NumMode_)) ' to ' num2str(RESDAMP(4,
                    NumMode_)) ' Hz'],...
228            ['Polynomial order ' num2str(RESDAMP(5,NumMode_))],...
229            ['Channel ' num2str(RESDAMP(6,NumMode_))]});
230    end
231
232    %% Changes settings after using the sliders
233    if strcmp(action,'updateslider')
234        V_=get(Slider_(1), 'Value');
235        AmpMax_=get(Slider_(2), 'Value');
236        normeigenvec_ = AmpMax_*MODE(:,NumMode_)/max(abs(MODE(:,NumMode_)));
237    end
238
239    %% Quit program
240    if strcmp(action,'stop')
241        RUN_=0;
242        close gcf
243    end
244
245    %% Stop and start animation
246    if strcmp(action,'RunOrStop')
247        if RUN_==1;
248        RUN_=0;
249        elseif RUN_==0;
250            animateData('animate')
251        end
252    end
253
254    %% Start and stop recording
255    if strcmp(action,'record')
256        if REC_==0;
257            set(Button_(3),'String','Stop')
258            REC_=0.5;
259        else
260            REC_=-1;
261            set(Button_(3),'String','Record')
262        end
263    end
264
265    end
```

Algorithm 3: Eigenmode animation

References

[1] Ulf Carlsson *Experimental Structure Dynamics*, KTH, 2016.

[2] Richard Markert *Strukturdynamik*, Shaker, 2013.

YOUR KNOWLEDGE HAS VALUE

- We will publish your bachelor's and
  master's thesis, essays and papers

- Your own eBook and book -
  sold worldwide in all relevant shops

- Earn money with each sale

Upload your text at www.GRIN.com
and publish for free